## Using Simple Machines

# Wedges All Around

by Trudy Becker

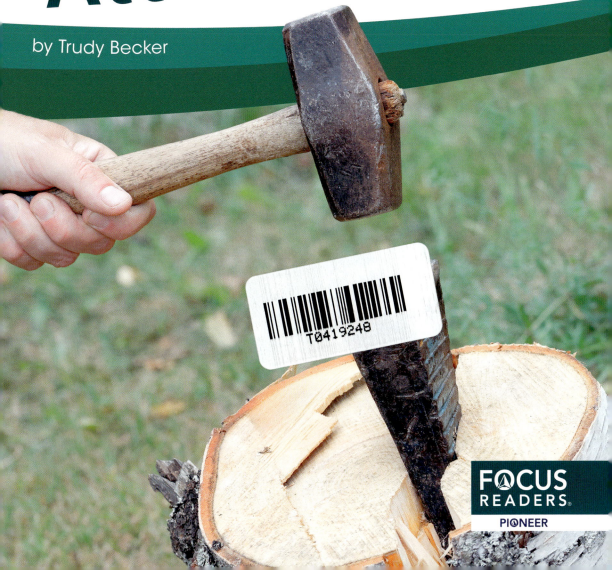

www.focusreaders.com

Copyright © 2024 by Focus Readers®, Lake Elmo, MN 55042. All rights reserved. No part of this book may be reproduced or utilized in any form or by any means without written permission from the publisher.

Focus Readers is distributed by North Star Editions:
sales@northstareditions.com | 888-417-0195

Produced for Focus Readers by Red Line Editorial.

Photographs ©: Shutterstock Images, cover, 1, 6, 10, 14 (bottom), 17, 18, 20; iStockphoto, 4, 8, 12, 14 (top)

**Library of Congress Cataloging-in-Publication Data**
Names: Becker, Trudy, author.
Title: Wedges all around / by Trudy Becker.
Description: Lake Elmo, MN : Focus Readers, [2024] | Series: Using simple machines | "Focus Readers Pioneer"-- cover. | Includes bibliographical references and index. | Audience: Grades K-1
Identifiers: LCCN 2023003203 (print) | LCCN 2023003204 (ebook) | ISBN 9781637396018 (hardcover) | ISBN 9781637396582 (paperback) | ISBN 9781637397701 (ebook pdf) | ISBN 9781637397152 (hosted ebook)
Subjects: LCSH: Wedges--Juvenile literature.
Classification: LCC TJ1201.W44 B43 2024  (print) | LCC TJ1201.W44  (ebook) | DDC 621.8--dc23/eng/20230208
LC record available at https://lccn.loc.gov/2023003203
LC ebook record available at https://lccn.loc.gov/2023003204

Printed in the United States of America
Mankato, MN
082023

# About the Author

Trudy Becker lives in Minneapolis, Minnesota. She likes exploring new places and loves anything involving books.

# Table of Contents

**CHAPTER 1**
## Swing and Chop 5

**CHAPTER 2**
## What Are Wedges? 9

**CHAPTER 3**
## Wedges Everywhere 13

THAT'S AMAZING!
## Making Art 16

**CHAPTER 4**
## Fun with Wedges 19

Focus on Wedges • 22
Glossary • 23
To Learn More • 24
Index • 24

## Chapter 1

# Swing and Chop

It is cold outside. A man wants to build a fire. He needs small pieces of wood. So, he gets his axe. He swings it down on a log. The log splits. Now the man can make a fire.

The man's axe has a wedge. Wedges are one of the six **simple machines**. Wedges can slice things. They can separate parts of an object.

**Fun Fact**

The heads of axes are wedges. Axe handles work like **levers**.

Chapter 2

# What Are Wedges?

All simple machines help people do jobs. People use wedges to break or cut things. But wedges can hold things in place, too. They can even hold different things together.

A wedge has a triangle shape. Two of the sides are **inclined planes**. These sides come together. They form a sharp edge. The edge pushes into things. That **force** makes wedges work.

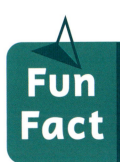

**Fun Fact** An inclined plane is another simple machine. It is shaped like a ramp.

## Chapter 3

# Wedges Everywhere

Wedges are all around. Knives are a common kind of wedge. Their sharp edges break things apart. Wedges are part of shovels, too. The wedge part breaks into the ground.

Some wedges hold things in place. A doorstop can hold a door open. Tacks can hold paper onto a board. Staples are a kind of wedge, too. They hold pages together.

**Fun Fact**: The point of a needle is a very thin wedge.

That's Amazing!

# Making Art

A **chisel** is a kind of wedge. People use chisels to chip away at things. Artists can use chisels to make **sculptures**. They can make sculptures out of wood. Or they can use ice. Artists can even use chisels to cut rocks.

Chapter 4

# Fun with Wedges

Wedges can be used for fun, too. Some people cook for fun. Cooks can use wedges such as cheese graters. Graters slice off bits of food. Pizza cutters have wedges, too.

Many kinds of **transportation** use wedges. Airplane wings cut through the air. The **bow** of a boat cuts through water. These wedges help people travel.

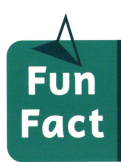

**Fun Fact**

Teeth are wedges. When people bite, their teeth split the food.

# FOCUS ON
# Wedges

*Write your answers on a separate piece of paper.*

1. Write a sentence that explains the main idea of Chapter 4.

2. What is the most helpful way you use wedges in your life? Why?

3. What other simple machine is part of a wedge?
    - A. a chisel
    - B. a cheese grater
    - C. an inclined plane

4. How does a doorstop hold doors open?
    - A. The wedge is very light.
    - B. The wedge is pushed tightly into place.
    - C. The wedge is bigger than the door.

*Answer key on page 24.*

# Glossary

**bow**
The front part of a boat.

**chisel**
A tool with a sharp edge used to carve hard materials.

**force**
A push or pull that changes how something moves.

**inclined planes**
Slanted surfaces, such as ramps. Inclined planes are one of the six simple machines.

**levers**
Bars that are used to move things. Levers are one of the six simple machines.

**sculptures**
Works of art that are shaped out of hard materials.

**simple machines**
Machines with only a few parts that make work easier.

**transportation**
Ways of moving people and goods from one place to another.

# To Learn More

## BOOKS

Blevins, Wiley. *Let's Find Wedges*. North Mankato, MN: Capstone Press, 2021.

Mattern, Joanne. *Wedges*. Minneapolis: Bellwether Media, 2020.

## NOTE TO EDUCATORS

Visit **www.focusreaders.com** to find lesson plans, activities, links, and other resources related to this title.

# Index

**A**
axe, 5, 7

**C**
chisel, 16

**D**
doorstop, 15

**I**
inclined planes, 11

Answer Key: 1. Answers will vary; 2. Answers will vary; 3. C; 4. B